36TH CONGRESS, 2d Session. } HOUSE OF REPRESENTATIVES. { Ex. Doc. No. 63.

ADDITIONAL ESTIMATE FOR FORT KEARNEY, SOUTH PASS, AND HONEY LAKE WAGON ROAD.

LETTER

FROM

THE ACTING SECRETARY OF THE INTERIOR,

TRANSMITTING

A communication from Colonel Lander in regard to the Fort Kearney, South Pass and Honey Lake wagon road.

FEBRUARY 11, 1861.—Referred to the Committee of Ways and Means, and ordered to be printed.

DEPARTMENT OF THE INTERIOR,
February 11, 1861.

SIR: I have the honor to enclose herewith, for the consideration of Congress, a communication from F. W. Lander, superintendent of the Fort Kearney, South Pass, and Honey Lake wagon road, and several petitions, numerously signed by emigrants, in reference to the construction of a bridge across Green river and some of the smaller streams in that vicinity and along said road.

There will remain about $10,000 balance of the appropriation of the above road applicable to this object; and I would respectfully call the attention of Congress to this work as being of great importance to the overland emigration, and recommend that the additional sum of $15,000 be appropriated to accomplish it.

I am, sir, respectfully, your obedient servant,

MOSES KELLY,
Acting Secretary.

Hon. WM. PENNINGTON,
Speaker of the House of Representatives.

WASHINGTON, *January* 5, 1861.

SIR: The subject of the construction of a bridge across Green river has been repeatedly referred to in my reports of previous years. In that of 1859 I related the fact of my stationing a party of men at that river, equipped with ropes and excellent mule teams, by the aid of which the emigrants crossed it without much difficulty, although some property was lost by them and one individual drowned. At that time the emigrants drew up two petitions asking that this river might be bridged, which, bearing several thousand signatures, in fact, the names of all the male individuals of their trains, were brought by me to Washington and referred to your department.

Last summer, while constructing the western division, I again met the emigration, and learned from it that, from the changes of the river bars, the ford had become nearly impracticable.

The emigrants argue, with much force, that after leaving the old road at South Pass they make several days' journey before reaching Green river. They have then either to attempt its passage, at much risk of property and life, or return to the ferry of the old road by an additional travel of two hundred miles. The force of the current over the sand bars at the new crossing is such as to preclude the establishment of a ferry.

The News Fork is another very dangerous and difficult crossing on the new road, as well as Smith's Fork of Bear river. If you should decide to expend the balance of the appropriation in the construction of a bridge at Green river, it would be expedient to add the sum of fifteen thousand dollars ($15,000) to the amount remaining of the appropriation and bridge the river below its junction with the News Fork. This sum would also enable a party, once equipped and in the field, to bridge the Smith's Fork of Bear river.

Your instructions of previous years have been so explicit on the subject of not erecting costly bridges, that I could not do more than lay these facts before you in my yearly reports. The view you have hitherto taken, that the bridges would be destroyed by the mountain traders owning ferries on the old roads, is undoubtedly a correct one; but if the important crossing of Green river is to be bridged, the expedition might be directed to erect there a common block-house and blacksmith's forge, and furnish them rent free to any reliable mountaineer of former expeditions. There are several individuals who, if thus provided for, would be glad to remain at the bridge and keep it in order for travel. I cannot apprehend the destruction of a properly constructed bridge by fire.

INDIANS.

It would be highly expedient to furnish the building party with at least one thousand dollars ($1,000) worth of Indian goods, to procure the further good behavior of the Washikee band of Snake Indians. The well known probity of Washikee would render any agreement made with him for the protection of the bridge a perfectly safe one.

Should you direct the expedition to then pass on towards California, which, regarding the sale of stock, would be the most economical course, I have the honor to again most urgently refer to my late report to the honorable Commissioner of Indian Affairs, detailing the causes and circumstances of the late Pah-Ute Indian war. By a conditional armistice concluded by me with the leading war chief of this tribe, hostilities were suspended by both the Indians and the emigrants and border whites for the space of one year. But, on the part of the Indians, it was with the express condition that I should lay the statement of the chief before the authorities at Washington, and use my best endeavors to procure some recognition of the claims of the tribe by government.

It will be impossible to again pass the unprotected emigration through this tribe, as I was enabled to do last summer, unless steps are taken to prove to the Indians the disposition of the government to notice their complaints. I have every reason to know that the Pah-Utes have, until very recently, been in league with the Shoshocos, and if I had possessed the authority to visit the latter as well as the Pah-Ute chief, that the massacre of emigrants, referred to in my report as having taken place near Salmon Falls, might have been prevented. While the Shoshocos held the upper road to Oregon, the Pah-Utes had assembled in large numbers along that to California, and were concentrating to attack the trains when the armistice was made.

Referring to my Indian report for further information, I would respectfully suggest that, if the expedition be ordered to California, such a portion of the Indian funds as you may deem expedient may be placed at the disposal of the officer in charge, that he may visit the Pah-Utes and lay before them your views on the subject of their application. Without wishing to intrude upon the province of the regular agent of this tribe, I have simply to say that, being thoroughly acquainted with their northern haunts and the points at which they usually assail emigrants, I should consider it a most cheerful duty to again visit them, and prevent their attacking trains.

Should you instruct me to do so, I have no doubt of being able to prevail on the principal chiefs to accompany me to Washington and execute a treaty here.

They desire to sell the lands now occupied by whites, or adjacent to the settlements, and thus create a fund out of which they can be taught to farm.

If these or other steps are taken I have reason to believe that the passage of the Pah-Ute Indian country will be practicable to emigrants, and that the war of western Utah will not be reopened by those savages at the close of the stipulated year.

The California road between Fort Hall and Tutt's Meadows on the Humboldt should either be protected by mounted rangers or by cavalry, directed to keep the field from June 15 to the middle of September.

This line of country is occupied by the Shoshocos or Western Snakes and Pannacks, during the passage of the emigrants, and the late trains are invariably attacked. The band of Snakes which frequents Salmon Falls on the Oregon road, during the fishing season, have numerous trails crossing the cañon country which divides Snake river from the Humboldt, *and direct their aggressions towards either road as opportunities offer*. No engagements made with these Indians can be regarded reliable until they are thoroughly chastised.

If the regular troops are not directed by the War Department to keep the field until all the trains have reached the Pah-Ute line, there will be no safety for emigrants while passing through the Snake country.

If it should be incompatible with your views to carry out so extensive a programme as is herewith submitted, or if Congress should fail to pass the additional appropriation, I have the honor to suggest that the balance of the funds now on hand be applied to the bridging of Green river, by a contract with some responsible party who will give bonds to keep the structure in repair, by being permitted to charge a low rate of toll, say one-tenth part of the price per wagon now paid by emigrants at the ferries of the old road.

Very respectfully, your obedient servant,

F. W. LANDER,
Superintendent.

Hon. JACOB THOMPSON,
Secretary of the Interior.

We, the undersigned, emigrants to California and Oregon, having just passed with our wagons and stock over the new government road from the South Pass to Fort Hall, (called Lander's Cut-Off,) do hereby state that the road is abundantly furnished with good grass, water, and fuel ; there is no *alkali* and no desert, as upon the old road, and while upon it our stock improved and rapidly recovered from sickness and lameness. We were much surprised at the great amount of labor that had been done in cutting out the timber and bridging and grading the road, and in all respects it more than met our expectations, especially those of us who have heretofore travelled the other routes. But we would most respectfully suggest that a bridge should be erected, as soon as possible, over Green river, the fording of which is dangerous and the cause of much trouble to the emigration, and in one instance the loss of life. We have been treated kindly and, in every case where the circumstances required it, aided and assisted on our way by the wagon road expedition, and we have likewise received the kindest treatment from the Indians, and we advise the overland emigration to California and Oregon to take this road, as the shortest and best adapted for the comforts of the traveller and the preservation of stock, especially if the government, in view of the many advantages of this route, should cause Green river to be bridged.

Statement of emigrants to California and Oregon.

Names and residence.	Number of wagons.	Number of persons.	Number of stock.	Place of destination.
Ferguson Chappell, Cedar county, Iowa	1	4	8	
S. H. Boardman, Connecticut	1	3	4	
J. M. Dewey, Wisconsin	1	3	6	Oregon
William Steel, Wisconsin	1			
James M. Torrence, Iowa	1	3	30	Oregon
Thomas Wilson, Illinois	1	5	4	do
R. H. Anderson, Missouri	12	41	240	California
Hiram Buell, Illinois	8	22	24	do
Joseph Woodward	9	40	115	do
William Fowler, Michigan	2	13	10	do
W. M. Orcult	3	9	14	
William Jacobs, Morris, Illinois	1	2	2	
Fayette Lincoln, Cook county, Illinois	3	11	11	California
J. R. & J. B. Adams	3	10	14	do
A. D. Gillson, Michigan	2	11	8	do
Daniel Hire	4	4	20	do
James Ferguson, New York	2	10	37	do
John P. Higgins, Illinois	1	5	8	do
Henry Robinson, Illinois	1	3	4	do
J. Holfreld, New York	1	8	4	do
William D. McIlroy, Iowa	14	37	209	Oregon
K. D. Todd, Iowa	1	4	7	do
Frederick Burhoff, Wisconsin	2	4	12	California
W. G. Brown, Wisconsin	8	30	48	do
Philip Hyde, Iowa	1	4	4	do
Jas. A. Archer, Iowa				
Stephen Rily, Iowa	1	3	6	California
A. M. Gaylard, Missouri	5	22	43	do
Peter Large, Illinois	3	14	11	do
William Black, Illinois	1	4	9	do
J. S. Perkins, Minnesota	2	4	8	do
A. Garrison, Iowa	4	4	18	do
I. A. Cardwell, Nebraska Territory	13	63	150	do
Dr. J. Goyer, Rock Island, Illinois	6	19	32	do
Robert Steer, Otsego, New York	3	17	19	do
S. H. C. Mason, Madison, New York	1	6	6	do
C. F. Kauffman, Oregon	3	15	35	Oregon
S. C. Jolly, Illinois	1	11	32	California
G. C. Ledyard, Illinois	1	4	17	do
J. S. Patterson, California	1	6	66	do
E. Crain, Missouri	1	4	17	do
F. H. Hathaway, Illinois	1	7	6	do
J. M. Coalter, Iowa		6	4	Oregon
R. E. Wood, Wisconsin	1	5	10	California
J. F. Lyon, Wisconsin	2	6	12	do
D. Vandehoof, Wisconsin	1	4	6	do
J. Brown, Wisconsin	4	5	14	do
William H. Sockrider	2	4	17	Oregon
A. J. Ball, Wisconsin	4	16	33	California
J. F. Wood, Wisconsin	6	23	54	Oregon
S. Maxson, Wisconsin	1	2	5	do
T. B. Borst, Iowa	3	11	17	California

STATEMENT—Continued.

Names and residence.	Number of wagons.	Number of persons.	Number of stock.	Place of destination.
William Coad, Illinois	4	16	22	California
Seth Ferrel, Iowa	1	2	4	Oregon
Brethnel Ferrel, Iowa	1	3	2	California
Caleb Witt, Tennessee	2	5	2	do
Jas. Witt, Tennessee	1	2	4	do
M. R. Renshaw, Iowa	1	4	5	do
William Coad, Iowa	1	3	7	do
J. D. Hanscom, Michigan	3	9	14	Oregon
Thomas Yunsal, Michigan	1	2	11	do
Seymour Carr, Iowa	1	3	13	California
George W. Newsom	14	53	103	Oregon
L. W. Dickey, Iowa	1	4	5	California
George Thaner, Iowa	2	4	11	Oregon
Bidwell Coons, Ohio	5	14	26	California
J. G. Smith, Missouri	1	4	6	do
N. B. Rine, Missouri	2	5	6	do
C. W. B., Ohio	1	2	7	do
Thomas Gunn, Missouri	10	52	370	do
A. J. Gallaway			16	do
William T. Ramsey, Augusta, Illinois	2	7	16	do
S. Ramsey, Augusta, Illinois	2	3	21	do
Theodore T. Ramsey, Augusta, Illinois				
Cawsun M. Rarnham, Wisconsin	1	4	11	Oregon
William E. Hoxie, Illinois (With W. T. Ramsey.)				
Charles Caldwell, Illinois (With W. T. Ramsey.)				
J. C. Rhoads, Illinois (With W. T. Ramsey.)				
Charles King, Illinois (With W. T. Ramsey.)				
S. M. Worthington, Kansas Territory	3	4	5	California
Napoleon P. Byrne, Missouri	7	23	225	do
Wheeler Elgin	3	18	474	do
William Rice, Independence, Missouri	8	37	879	do
David Cummins	1	3	25	do
Daniel Powell, Illinois	1	3	10	Oregon
J. Adams, captain, Lagrange county, Indiana company	4	12	43	California
P. H. Poindexter, California	8	30	300	do
Stephen O. Gray, Michigan				
Charles H. Conklin, Michigan	1	7	2	California
G. W. Winder	1	7	16	
H. Tuel, Iowa	1	4	21	California
S. Gilliland, Wisconsin	1	2	3	do
O. B. Nellis, Michigan	2	9	14	do
Levi R. Geer	1	3	5	do
James McClosky, Michigan	1	4	4	do
Bela Rathbun	9	42	56	Oregon
S. P. Jallen	1	4	4	
F. M. Rice, Iowa	5	21	320	California
A. Vangiesen, Canada West	3	12	16	do
Sobieski Brown, Canada West	3	9	17	do
Elmore J. Ferguson, Illinois	3	5	7	do
Job Huff, Nebraska Territory	1	3	2	do
T. Wallingbock	1	7	7	do

STATEMENT—Continued.

Names and residence.	Number of wagons.	Number of persons.	Number of stock.	Place of destination.
Isaac Walker, Iowa	1	4	7	California
Richard Gant	....	....	....	....
Sam T. Welch	3	12	85	California
C. B. Welch	4	11	29	....do
William Haskin	2	4	6	....do
E. Griffith, Iowa	2	5	8	....do
A. H. Whitcomb, Illinois	2	6	7	....do
Edwin Pett, Illinois	1	3	2	....do
J. Camton, Illinois	2	4	8	....do
W. Townsend, Illinois	1	2	4	....do
H. Whipple, Michigan	1	2	11	....do
E. W. Mahoney, Iowa	2	4	6	....do
H. Thompson, Illinois	1	2	4	....do
W. Thompson, Illinois	3	5	14	....do
William A. Allard, Illinois	1	4	4	....do
M. J. Sampson, Illinois	2	11	8	....do
O. F. Sampson, Illinois	....	....	....	....
Joseph Cheasebro, Illinois	1	2	2	Oregon
Sherman Hatch, Illinois	3	5	15	California
C. D. Needham, Illinois	....	....	....	....do
Stanley Willey, Iowa	1	4	8	....do
John Hoy, Ohio	1	4	16	....do
John Wells, Iowa	4	14	40	....do
George W. Harvey, Ohio	....	....	....	....do
Franklin Connelly, Iowa	....	....	....	....do
Andrew Tash, Iowa	....	....	....	....do
Bedy Akers, Iowa	....	....	....	....do
Thomas Ralston, Iowa	....	....	....	....
F. Noble, Iowa	....	....	....	California
Jonathan Boyce, Iowa	3	14	38	....do
William C. Humphrey, Illinois	2	7	15	....do
J. W. Wilson	....	....	....	....
William E. Harper	1	2	5	California
J. T. Vaughan, Iowa	....	1	8	....do
M. A. Babcook, Iowa	2	7	19	....
James A. Smith, Iowa	1	6	11	California
A. Babcook, Iowa	1	4	16	....do
E. Babcook, Iowa	....	1	31	....do
J. F. Knowles, Iowa	....	6	11	....
J. W. Babcook, Iowa	2	7	47	California
George Urie, Iowa	3	....	....	Oregon
W. Babcook	2	12	26	....do
C. A. Daniels, Michigan	1	7	9	....
Jesse Perkins, Michigan	2	7	15	Oregon
Salem Longeon	1	3	7	....
J. McGinnis	1	8	1	....
Peter Saling	3	7	49	Oregon
Thomas McKee, Michigan	1	4	4	....do
Wm. C. LaDow, Iowa	1	1	1	California
O. W. Kelley, Iowa	....	2	3	....do
S. W. Fero, Iowa	....	1	2	....
Thomas Nuttle	....	2	4	....

STATEMENT—Continued.

Names and residence.	Number of wagons.	Number of persons.	Number of stock.	Place of destination.
Thomas Brady	1	2	4	
E. P. Wirts	1	5	8	California
A. Ashworth, Michigan		1		do
J. L. Davies				
D. H. Carpenter, Iowa				California
John R. McClure	1	4	4	do
John L. Davis	3	10	36	do
Robert Hastie	4	7	23	do
P. Farlie	1	4	40	do
William Moore, Virginia	4	10	34	Oregon
T. P. Denney	4	10	65	do
F. S. Turner	1	2	2	do
A. T. Shreeves	3	13	17	do
G. Hutchinson, Iowa	1	4	5	do
William W. McHenry, Iowa			2	do
M. Welch, Ohio	2	9	17	do
D. Houck, Ohio			1	
E. N. Dunbar, Indiana	1	4	7	California
H. M. Kingsbury, Ohio	2	11	13	do
Edwin Houghton	2	8	17	do
O. S. Coddington, Indiana			3	
O. M. Jackman	2	7	13	California
D. J. Houghton, Ohio				
F. Bellup, Ohio	13	22	78	California
John Price, Ohio	3	7	19	do
D. Best, Ohio	1	3	11	do
D. Yerby, Iowa	1	4	13	Oregon
H. Billups, Iowa	4	9	17	do
E. Hemingway, Virginia	11	19	49	California
Thomas J. Bunker, Virginia		1	2	do
E. C. Lindsey, Oregon	1	2	13	do
Abram Lindsay, Ohio	3	7	19	do
Thomas Lindsay, Ohio	1	4	11	
W. C. Adams, Ohio	1	1	7	California
L. Stockman, Ohio		1	3	Oregon
Joseph M. Bock				
H. H. Hill, Wisconsin	1	2	9	California
S. F. Ledyard, Missouri	2	7	19	
J. W. Burnell, Maine		1	1	
Marshel Murray, Illinois	2	9	17	
James Cheney, Illinois	7	11	34	Oregon
Thomas Sulivan, Illinois	1	3	11	do
James Larninger, Michigan		1	1	
William Reynolds, Indiana	9	13	47	California
M. Alberron, Indiana	3	11	23	do
S. Preston, Missouri	3	7	17	do
John Moore, Missouri		1	1	
E. Phelps, Missouri	13	22	117	Oregon
J. Leinenger, Missouri	4	13	79	do
B. Doty		1	3	California
R. S. Wilkin, Indiana	7	19	85	do
C. Burget, Indiana	1	3	17	Oregon
J. E. Carrol		1	4	California

STATEMENT—Continued.

Names and residence.	Number of wagons.	Number of persons.	Number of stock.	Place of destination.
J. S. Waldridge, Illinois	1	4	4	California
D. E. Knight, Illinois	2	9	11	do
W. Brown, Illinois	4	8	18	Oregon
A. F. Wells, Michigan	1	3	5	
John Rice				California
Thomas Lewis, Indiana	1	2	3	do
E S. McClellan, Michigan		1	1	do
Ambers Thornburgh		1	9	Oregon
John Phelan, Iowa	3	11	13	do
Emmor Ramsey				
A. J. Clark, Indiana				California
C. Carson	11	17	49	Oregon
D. R. Bittinger, Michigan	2	7	9	
James Stone, Ohio	1	5	7	California
J. Warley, Ohio	2	7	17	do
S. W. Puffingher, Indiana	7	9	29	Oregon
Wm. Bradford, Wilmington, Del		1	3	California
George Robinson, Iowa		13	37	do
Peter Helbey, Virginia				
Amos Barnhart	7	14	39	California
S. C. Burns, Virginia		1	1	
B. Manning, Ohio	11	17	48	California
John Woods, Ohio				
H. Herman, Iowa	2	5	13	California
F. M. Jolly, Iowa	1	6	48	do
F. M. Mounts, Iowa	1	5	11	Oregon
J. E. Moore, Iowa	1	4	17	do
E. G. Banks, Michigan	3	11	43	do
Henry Burket, Ohio	13	19	91	California
N. Piles, Ohio	2	9	18	do
George Bradfield, Ohio	9	18	89	do
F. M. Lewis, Indiana		1	7	
T. F. Ryan, Indiana	19	31	211	Oregon
David Carter, Indiana				do
David Enos		1	1	California
Wm. Moore, Ohio	11	17	129	Oregon
Daniel Claton				
John M. Bryan, Ohio	2	11	39	
J. T. Hartman, Virginia	7	19	229	California
William Burgett				
A. F. Core	2	11		California
E. R. Wright, Iowa	2	8	23	do
Edwin G. Wood				
E. E. Davies, Iowa	7	19	47	California
John Marshall				
Frank Demask		1	2	Oregon
George Wood, Ohio	2	8	17	do
Lanis Darling, Ohio	13	32	273	do
O. Olsan, Oregon	1	5	12	do
Geo. Klingaman		1	3	
Thomas McVay, Oregon	7	19	123	do
Warren Dunham, Ohio	2	13	27	California

STATEMENT—Continued.

Names and residence.	Number of wagons.	Number of persons.	Number of stock.	Place of destination.
Alexander Glen, Indiana	1	5	23	California
J. L. C. Sherwin, California	5	14	32	Oregon
Wm. Adle, Illinois	2	6	15	do
John Walling, Illinois	1	7	19	do
Franklin Shores, Illinois	3	11	39	California
Samuel W. Empey, Michigan	7	19	73	
Charles Helm, Wisconsin	3	11	24	California
John F. Bush, Wisconsin	1	3	11	
James N. Anderson, Missouri	13	23	83	Oregon
Ira A. Garrison, Wisconsin				do
E. Fifield, Maine		1	7	California
J. L. Thompson, Wisconsin	7	19	41	do
G. W. Brown, Wisconsin	1	3	9	do
David Brown, Wisconsin		1	3	do
Lewis S. Kelsey, Missouri	19	29	108	Oregon
John Beadle, Iowa		1	1	California
Park Winans, Illinois	1	2	3	do
Geo. W. Grimes, Iowa	2	9	11	do
Emanuel Brannan			7	do
Frederick Morrisson, Missouri	1	3		
Claiborne Vaughan, Indiana	1	3	11	Oregon
S. F. Lewis, Indiana	2	7	8	do
J. H. Claughton, Missouri	1	4	10	California
W. B. Wooldridge, K. T	1	9	15	Oregon
George Peck, Minnesota	1	5	8	California
John Fr. Adler	7	14	98	do
Samuel Allen	1	3	6	do
E. A. Hall		1		
Albert Allen	1	3	7	California
Reuben Allen, Minnesota		4	5	do
S. S. C. Spencer, Minnesota	4	11	61	do
F. Kingman, Illinois	1	6	11	do
Charles Kingman, Illinois	3	9	49	
Daniel A. Ellis, Illinois	1	4	8	Oregon
Wm. Albaugh, Iowa	1	4	8	do
Jacob Gauntz, Iowa	1	4	6	California
James Runyan	3	7	19	
W. L. Mathews, Iowa		1	2	California
Jas. H. Underwood		1	1	Oregon
Alfred Graham, Indiana	11	17	119	do
Thomas Brown, Indiana	1	4	5	California
James H. Story, Iowa		1	1	Oregon
Wm. O. Miller	5	15	30	California
Andrew H. Dennick	1	2	4	do
Henry J. Miller	1	9	11	Oregon
John Bryan		1	1	
J. Amoore, Maine	17	28	117	Oregon
Wm. Fee, Missouri	3	13	48	
W. Sherwin, Missouri		1	5	California
Geo. Jones, Virginia	19	31	238	Oregon
H. A. Leavens, Illinois	5	20	45	do
J. S. Beamis, Illinois	7	13	27	do

STATEMENT—Continued.

Names and residence.	Number of wagons.	Number of persons.	Number of stock.	Place of destination.
T. Dickerson, Illinois	1	5	13	Oregon
H. H. Case, Wisconsin	17	27	117	do
Thos. S. Sloane, Wisconsin	3	9	17	do
Alfred Hawk		1	3	do
Chs. Fegley, Iowa	13	17	73	California
S. J. Dickerson, Iowa	7	11	39	do
Gregor Shreeve	1	4	9	do
Hiram Cain				
F. T. Howard	1	3	3	California
Geo. W. Howard		1	2	do
H. C. Minick	3	7	43	do
Robert Schenk, Wisconsin	7	13	79	Oregon
Joseph Britt, Iowa	1	4	9	do
S. H. Hinds, Wisconsin	2	7	17	do
Richard Talbart, Illinois		1	2	do
W. M. Franks, Illinois				
James H. Bowen, Illinois	1	3	23	California
F. Michaelson & Co., Illinois	5	15	40	
T. Michaelson, Illinois		1	1	
Anderson, Illinois				
G. Michelson, Illinois	2	5	17	California
H. Stak, Iowa	7	19	75	do
Lauson, Iowa		1	2	do
H. Seeman, Iowa	13	27	276	California
A. Seeman, Iowa				
P. Seeman, Iowa				
T. Seeman, Iowa				
T. Streiff, Iowa				
T. Streiff, Iowa				
H. Hunt, Iowa				
T. Serringer, Iowa				
T. Boilsle, Iowa				
F. Lagerson, California	6	27	46	
Peter Timm, California	3	17	171	California
Peter Jacobs, California				
Martin Haman, California				
Marx Dittmer, California				
Peter Sanger, California				
Johann Trede, California				
Claus Gergen, California				
John Dussler, California				
Peter Carsten, California				
Claus Hinrichsen, California				
John Rohneer, California				
Henri Weinholz, California				
Friderich Frahm, California				
Henri Soos, California				
John Soos, California				
Henri Timm, California				
Henri Thomsen, California				
Christian Leisner, California				
John Miller, California				

STATEMENT—Continued.

Names and residence.	Number of wagons.	Number of persons.	Number of stock.	Place of destination.
Peter Conk, California				
Claus Budendorf, California				
Joachim Saga, California				
Charles Willmaka, California				
Joseph Watson	13	29	269	Oregon
Lewis M. Garvin	7	18	73	California
Lancelot Carr	5	15	36	do
Benjamin J. Curler	4	18	35	do
William Gughton, Illinois	7	15	24	do

STATE OF CALIFORNIA, *City and County of San Francisco, ss.*

E. P. Ream, being duly sworn, deposes and says that he is acquainted with H. Seemann, whose name is written at the head of the list of names on this sheet; that the said Seemann acknowledged to this deponent that this is his genuine signature, and also that he wrote the names following his at the request of the said parties thereto, who were members of a train of emigrants under his lead, bound to California.

E. P. REAM.

Subscribed and sworn to before me October 25, 1859.

HENRY HAIGT, *Notary Public.*

FORT HALL,
Oregon Territory, July 15, 1858.

This is to certify that we, the undersigned, have travelled over the Pacific wagon road, better known as Lander's Cut-Off, and find it a very acceptable road for emigrants. We think it preferable to any other road across the mountains in many respects; most of the way it is well worked, and with a bridge across Green river (the only stream at all troublesome) it would be as good a road as many now travelled in the States; it is some five days' travel shorter than any other road across the mountains. There is no desert to cross on this route, no alkali to kill your stock; but instead, plenty of good water, abundance of grass, and wood enough to satisfy any reasonable man. Many of the undersigned have crossed by other routes, and give this the preference.

William Glaze, Missouri.
J. B. Nevins, New Hampshire.
Erastus Downing, Missouri.
William Martin, Missouri.
William A. Stone, Missouri.
Ole. Emins, Wisconsin.
William Flanagin, Missouri.

George W. Linderman, Illinois.
George W. Brown, Iowa.
Alexander Anthony, Missouri.
John Bagby, Missouri.
Marion Stow, Missouri.
E. W. Newkerk, Iowa.
A. Clubb, Iowa.
B. F. Griswold, Iowa.
Henry Y. Goldsmith, Wisconsin.
F. Williams, Wisconsin.
Joseph M. Nelson, Ohio.
Jervies J. Hedgpeth, Missouri.
William Wright, Missouri.
Jacob Arter, Iowa.
Amos Crater, Iowa.
James S. Mooney, New Hampshire.
William Norman, Illinois.
Thomas Redy, Illinois.
John Longhead, New York.
D. S. Sage, Wisconsin.
James Wetherhead, Wisconsion.
William Carll, Illinois.
John Wetherhead, Wisconsin.
Joseph G. Daniel, Iowa.
Amos Smith, Wisconsin.
Thomas Butler, Wisconsin.
David Chubb, Wisconsin.
James Contell, Wisconsin.
Charles Kaye, Wisconsin.
William Shirly, Wisconsin.
David Atkain, Wisconsin.
William Robertson, Wisconsin.
Thomas K. Ober, Wisconsin.
Luke Smith, Wisconsin.
Lyman Carpenter, Iowa.
George W. Martin, Iowa.
Garret Clawson, Iowa.
S. T. Armstrong, Trenton, Wisconsin.
Geo. Gray, Nininger City, Minnesota.
G. R. Kidder, Claremont, Minnesota.
Allen Mead, Illinois.
Robt. Steere, Oswego county, New York.
G. W. Squires, Carl county, Illinois.
C. F. Kauffman, Louisa county, Iowa.
A. H. Kauffman, Louisa county, Iowa.
E. R. Wood, Palmyra, Wisconsin.
R. E. Woods, Omaha City.
John H. Squier, Cass county, Michigan.
Wm. Wheeling, Bernadotte, Fulton county, Illinois.
A. W. Robinson, Bernadotte, Fulton county, Illinois.
J. R. Carey, Illinois.

Hill Burkhart, Washtenaw, Michigan.
J. G. Smith, Missouri.
N. B. Rine, Missouri.
J. T. Day, Missouri.
J. M. Kaufman, Missouri.
E. Crain.
Charles Lawrence.
J. L. Kinkade.
Charles Jolly.
W. H. Wells, Wisconsin.
J. M. Contter.
E. H. Scott.
S. C. Whitlatch, Illinois.
F. H. Hathaway, Illinois.
James Patterson.
O. J. Rogers.
R. Christy.
Martin Christy.
Michael Bourk, Wisconsin.
Amberson Huff, Michigan.
George McVicar, Winconsin.
R. W. Tilton, Washington county, Pennsylvania.
C. Hickey, Wisconsin.
G. R. Vansiclen, Michigan.
James Guild, Chicago, Illinois.
John A. Hickey, Wisconsin.
John Pettinger, Dubuque, Iowa.
Benjamin Sanders, Marengo, Illinois.
Howard Peterzon, Minnesota.
George Quigle, Illinois.
Alfred Graff, Elgin, Kane county, Illinois.
J. S. Deneson, Michigan.
Chester Smock, Minnesota.
H. B. Beach, Marengo, Illinois.
John Quigle, DeKalb, Illinois.
John G. Sneider, Anderson county, Kansas.
Jacob Whitbeck, Delhi, Iowa.
Edwin G. Kinne, Oconomowock, Wisconsin.
Francis Eaton, New York.
Charles W. Ryder, Wisconsin.
J. H. Ingersoll, Delhi, Iowa.
Thomas Eagan, Waukesha, Wisconsin.
Peter Eagan, Waukesha, Wisconsin.
James Johnson, Illinois.
Charles Follansbee, Kane county, Illinois.
James E. Beach, Elgin county, Illinois.
Riley McHenry, Elgin county, Illinois.
Amos Van Vleck, Wisconsin.
Wm. H. Springer, Michigan.
John Arnold, Wisconsin.
William B. Tiffany, Hastings, Minnesota.

Salmon Scott, Oakland, Michigan.
Bartlett A. Day, Minnesota.
George W. Springe, [illegible.]
James E. Harvey, Minnesota.
Reuben Burroughs, Ontario county, New York.
James Coughran, Reedsburg, Wisconsin.
Comfort H. Knapp.
Thomas V. R. Rathbun.
Samuel Coughran, Reedsburg, Wisconsin.
T. H. Jewett.
Henry B. Gaige, Reedsburg, Sauk county, Wisconsin.
James W. Beebe.
Albert Marston.
Levi S. Reed.
John A. Bloomer.
David C. Reed.
James Law.
George Winchester.
Leonard Low, Iowa.
John W. Allen, New York.
Jerome Beebe.
John Anderson, Iowa.
G. W. Colwell, Iowa.
William A. Evans, Pennsylvania.
John Levander, Iowa.
Charles Catterell, Iowa county, Wisconsin.
Levi Burgess, Wisconsin.
Anthony James, Wisconsin.
George Selvester, Wisconsin.
Abraham Selvester, Wisconsin.
W. H. Legol, Wisconsin.
George W. Hill, DeKalb county, Illinois.
James Temple, Illinois.
D. C. Adams, Iowa.
Thomas Walker, Wisconsin.
A. M. Johnson, Iowa.
John Stanaway, Wisconsin.
S. S. Chandler, Wisconsin.
H. H. Rice, Wisconsin.
R. Alderson, Missouri.
E. C. Sessions, Wisconsin.
B. F. Saltzman, Wisconsin.
A. C. Coates, Wisconsin.
Thomas J. Coates, Wisconsin.
Charles P. Traber, Wisconsin.
Jesse L. Coates, Wisconsin.
M. D. Dyhee, Kentucky.
Thomas Anderson, Kansas Territory.
Christian Finger, Kansas Territory.
Thames G. Murray, Kansas.
E. B. Purdom, Franklin, Kansas.

J. Bowley, Lawrence, Kansas.
Moses Wright, Indiana.
M. H. Merton.
J. H. Bennett.
Harry Burk.
J. C. Purdom, Franklin, Kansas Territory.
Benjamin Purratt, Franklin, Kansas Territory.
David Vanostan, Franklin, Kansas Territory.
James Roggers, Franklin, Kansas Territory.
A. H. Earby.
H. W. Tiel.
S. W. Smith, Freeport, Illinois.
S. Gregory, Blackford, Indiana.
H. M. Wells, Illinois.
R. Haines, Jackson, Iowa.
William Rice, Jackson, Missouri.
John C. Richardson, Janesville, Wisconsin.
Charles P. Murphy, Janesville, Wisconsin.
Francis Gafferty.
John G. Alason, Iowa.
Isaac Bradwell, Alleghany county, Pittsburg, Pennsylvania.
C. L. Lamoreux, La Porte, Indiana.
Charles A. Sankey, Pittsburg, Pennsylvania.
Almon Menter, Homer, New York.
Martin Menter, Syracuse, New York.
John Thornbury, New Cumberland, Virginia.
D. B. Conger, Homer, New York.
C. Slack, Michigan.
G. M. Pierce, Michigan.
William Walton, Pennsylvania.
James Brenneman, Ohio.
Chester Menter, Homer, New York.
John McMichael, Alleghany county, Pennsylvania.
John H. Sawyer, Bristol, Kenosha county, Wisconsin.
G. B. Franklin, Fort Dodge, Iowa.
P. McVicar, Salem, Wisconsin.
R. Spencer, Bristol, Wisconsin.
John W. Cleveland, Bristol, Kenosha county, Wisconsin.
O. S. Smith, Kossuth county, Iowa.
Dr. Joel Richardson, Hartland, Maine.
Benjamin C. Berwise.
Dr. O. W. K. McAllister, Blue Earth county, Minnesota.
Alfred N. Ludington, Dallas county, Iowa.
James Manitz, Freeport, Illinois.
Charles L. Buckman, Franklin, DeKalb county, Illinois.
Augustus Stiger, Freeport, Illinois.
Joseph Schwab, Freeport, Illinois.
Frederick Stoll, Freeport, Illinois.
O. L. Manfield, Warren, Illinois.
C. P. Ludesher, Davenport, Iowa.
George S. Lamin, Freeport, Illinois.

D. T. Culbertson, Wisconsin.
D. D. Atkinson, Wisconsin.
H. H. Longley, Rochester, Minnesota.
Edward K. Finson, Iowa.
Alexander Spaulding, Minnesota.
R. Sherer, Minnesota.
Rufus Emery, Maine.
Homer L. Clark, Iowa.
William Todd.
Ferguson Chappell, Cedar county, Iowa.
Nicholas Simmons.
Franklin Finson, Iowa.
Moses H. Finson, Iowa.
Rufus B. Emery.
R. N. McCollum, Michigan.
S. S. Cox, Liberty, Michigan.
George W. McCollum, Michigan.
James Young, Michigan.
Caswell Coil, Logan county, Illinois.
Philip Marvel, Missouri.
William Black, Mobile, Alabama.
C. C. Parker, Springfield, Illinois.
George C. Hurd, Menasha, Wisconsin.
A. D. Nelson, Jackson county, Michigan.
E. P. Bradford, Cook county, Illinois.
William Jacobs, Grundy county, Illinois.
[One name illegible.]
John P. Higgins, Richland county, Illinois.
Elisha Swift, Hillsdale, Michigan.
William Fowler.
John Millikan, North Carolina.
L. H. Rouze, Ohio.
Samuel Dagget, Mercer county, Illinois.
James Ferguson, Brownville, New York.
Almiron Dagget, Warren county, Illinois.
E. D. Ketchum, De Kalb county, Illinois.
John Case, Iowa.
Richard Jones, Illinois.
Fayette Lincoln, Cook county, Illinois.
David Wilson, Marion county, Iowa.
D. C. McKercher, Stephenson county, Illinois.
Joseph Woodward.
John C. Creswell, Multnomah county, Oregon.
James Cummings, Fremont county, Iowa.
E. Humphrey, Illinois.
William Orcutt, Hillsdale county, Michigan.
[One name illegible.]
James Daily, Michigan.
[One name illegible.]
A. J. Clem, Iowa.
Joseph Whaling, Illinois.

J. L. Wagner, Illinois.
Oliver Bowers, Illinois.

We, the undersigned, emigrants to California and Oregon, having just passed, with our wagons and stock, over the new government road from the South Pass to Fort Hall, (called Lander's Cut-Off,) do hereby state that the road is abundantly furnished with good grass, water, and fuel; there is no alkali and no desert, as upon the old road, and while upon it our stock improved and rapidly recovered from sickness and lameness.

We were much surprised at the great amount of labor that has been done in cutting out the timber and bridging and grading the road, and in all respects it more than met our expectations, especially those of us who have heretofore travelled the other routes; but we would respectfully suggest that a bridge should be erected as soon as possible over Green river, the fording of which is dangerous and the cause of much trouble to the emigration, and in one instance the loss of life.

We have been treated kindly and, in every case where the circumstances required it, aided and assisted on our way by the wagon road expedition, and we have likewise received the kindest treatment from the Indians; and we advise the overland emigration to California and Oregon to take this road, as the shortest and best adapted for the comfort of the traveller and the preservation of stock, especially if the government, in view of the advantages of this route, should cause Green river to be bridged.

Statement of emigrants to California and Oregon.

Names and residence.	Number of wagons.	Number of persons.	Number of stock.	Place of destination.
C. J. Bullock, Illinois	1	2	9	
Ervin Crane, Michigan	2	5	12	Oregon
O. J. Britton, Missouri	1	2	4	....do
John Monholland, Missouri	1	4	6	California
Franklin Durshee, Iowa	1	3	8	....do
Richard Dowler	1	2	9	Oregon
C. I. Peterson				....do
A. Cullings	2	9	17	California
——— Marvel, Missouri	1	2	4	....do
John F. Moore, Kansas	1	7	12	....do
Caswell Coil, Illinois	1	7	8	....do
James Young, Illinois	1	3	6	....do
James Caruthers, Wisconsin, (1 pack mule)	1	5	7	....do
William Jones, Missouri	2	9	15	....do
Eli S. Newson	1	4	10	....do
Cyrus Crouch	1	1	2	....do
Thomas Goodhue, Illinois	1	1	3	....do
D. Loveland, (passenger)				....do

STATEMENT—Continued.

Names and residence.	Number of wagons.	Number of persons.	Number of stock.	Place of destination.
Allen Ensley	3	5	28	California
German Buckland, Illinois	4	8	23	do
John W. Ensley	1	2	4	do
C. E. Parker, Springfield, Illinois	4	8	27	do
Truman C. Clark, Illinois	1	2	8	do
A. M. Gibbans, Illinois	1	3	2	do
F. G. Gilbert, Illinois	1	3	2	do
R. C. Maynes, Illinois	3	13	33	do
L. Klerkx, Illinois				do
J. D. Freeman, Minnesota				do
George A. Bronson, Iowa				do
R. Layton, Iowa				do
John Poplman, Wisconsin				do
E. Bronson, Iowa				do
Johann Blum, Wisconsin				do
Amasa Adams, Michigan				do
G. H. Brown, Illinois				do
J. S. Brown, Iowa				do
Martin Dean, Wisconsin				
L. B. Barkalow, Iowa				California
H. W. Rumze				
E. G. Rumsey				
W. B. McCune				
T. C. Tenwick				
S. Newell				
J. S. Dodds				
S. B. Carr				
Wm. Pursel				
Wm. H. Payn				
J. Mitchell				
Marcus Tenwick				
John Batgen				
John Johnston, Illinois				California
Wm Cherry, Illinois				do
Oliver Johnston, Illinois				do
Wm. Isirel, Missouri				do
James Isirel, Missouri				
Leman G. Hall, Illinois				
Alex. Carpenter				
P. W. Cunningham				
M. Bell				
John R. Lam				
T. H. Ekley, Michigan	1	2	4	California
N. Coffin, Wisconsin	1	4	8	
S. B. Butler, Wisconsin	1	2	4	
Robert Duffey, Illinois				California
Wm. Taylor, Illinois				do
M. R. Croft, Iowa				do
B. F. Couch, Pennsylvania				do
H. White, Illinois				do
Edwin C. Marshall, Illinois				do
N. R. Penney, Illinois				do

STATEMENT—Continued.

Names and residence.	Number of wagons.	Number of persons.	Number of stock.	Place of destination.
P. B. Lovett, Illinois				California
G. W. Buck				
L. H. Brady				
John Walter, Illinois		1		California
John Stewart, Wisconsin				
Robert Steen, Wisconsin		1		California
James McNaughton, Wisconsin		1		
[One name illegible]				
Samuel McNaughton, Wisconsin				
Malcolm McNaughton, Wisconsin		1		California
Alexander Vass, Wisconsin				
Charles Kittleson				
And. W. Calley	3	18	24	Oregon
John Foy				
Rufus Ammon				
Robert Rintoul				
Robert McCalley				
George Reining, Ohio	4	25	35	Oregon
Sylvester Patten, Wisconsin	4	25	35	do
Lester Patten, Wisconsin	4	25	35	do
W. G. Nickerson, Wisconsin		25	35	do
Wm. Babcock, Wisconsin	4	25	35	do
A. Sconton, Wisconsin		25	35	do
John Fronk			35	do
D. S. Bonsom, Wisconsin			35	do
F. Homes, Ohio	4	25	35	do
G. W. Wolder, Illinois	4	25	35	do
Mark M. Powell, Wisconsin	4	25	35	do
Edward Allen, Wisconsin				
George Lonel				
John R. Benefiel, Indiana				
John Thomson, Illinois				Oregon
Tristram Mayhew, Massachusetts				do
D. W. Harris, Illinois				
Thomas Walters, Illinois	4	25	35	Oregon
Charles Porter, Illinois				
H. Parker				
Luther Hower				
Ransom Northup				
Ira Isted, Iowa				
Aiken Tart, Iowa				
Henry M. Hall, Michigan	7	15	38	Yreka, Cal
G. B. Ross, Wisconsin				do
John Johnston, Illinois				California
J. M. Gilliland, Illinois				do
W. F. Everett, Illinois				do
William Gill, Illinois				do
Alfred J. Cooper, Illinois				do
George W. Scofield, Illinois				do
Thomas Fuller, Illinois				do
William Hill				
R. D. Harkness, Illinois				California

STATEMENT—Continued.

Names and residence.	Number of wagons.	Number of persons.	Number of stock.	Place of destination.
S. H. Loomis, Illinois				California
H. S. Phillips, Illinois				do
Russell J. Wells, Iowa	1	4		Oregon
George S. Barnes, Wisconsin	1	4		do
Hiram M. Jones, Iowa	3	15		California
Michael Hallasy, Illinois	2			do
Ira Moshier, Iowa				do
George Sharp, Iowa				do
J. R. Bennett, Indiana				do
J. N. Bennett, Indiana				do
George Clark, Michigan				do
James Ardery				
Chapman Warfins				
Patrick Age				
Edward Kerr				
Michael Fegan				
Daniel Matthews, Illinois				
Edward Fagan, Illinois				
Patrick Deny, Illinois				California
Robert Perry, Illinois				do
John Kane, Illinois				do
James Sinaman, Illinois				
Jared Kerble				
Thomas E. Stanton, Iowa	6	21	35	California
Elon White				
F. H. Hazard				
Joel Stanton				
A. Waterman				
Edward Pew, civil engineer				
N. R. Stanton				
Robert Brown				
William Bunting				
Joseph Richardson				
Hiram Dodge, Neb	4	14	25	Oregon
George W. Stone				
Augustus W. Burrill, Maine	1	4	8	Oregon
W. J. Broughton			6	
George W. Crist, Iowa				California
H. R. Dickerson, Illinois				do
E. A. Dodge, Illinois				do
P. Lightle, Illinois			9	do
R. F. Brown, Illinois				do
H. E. Rankin, Illinois				do
H. H. Lewis, Nebraska Territory				do
Rut. Hoffman, Ohio				do
J. B. Hoffman, Ohio				do
H S. Tarrington, Illinois				do
John Balluf, Illinois				do
W. Parsons, Iowa				Oregon
J. A. Everett, Illinois				California
H. Harrington, Illinois				do
A. La Veille, Illinois				do

STATEMENT—Continued.

Names and residence.	Number of wagons.	Number of persons.	Number of stock.	Place of destination.
John Kunkle, Illinois				California
John W. Balluf, Illinois				...do
Joseph Balluf, Illinois				...do
A. B. Moore, Iowa	5	19	30	Oregon
J. H. Dow, Iowa	4	12	64	...do
William Richardson, Iowa	2	5	11	...do
R. S. Butler, Iowa	1	2	2	...do
J. Bierschback, Wisconsin	2	6	6	...do
L. Kords, Wisconsin		6	6	...do
F. Kords, Wisconsin		6	6	...do
E. G. Crane, Illinois	2	7	8	California
Sylvester Critz, Iowa	1	3	5	...do
S. W. Maxwell, Illinois	6	20	63	...do
James R. Hyhes, Illinois	7	19	90	...do
Samuel A. Bone, Illinois	2	9	16	...do
J. J. Kunzler, Illinois	1	4	6	...do
Alexander Brander, Illinois	1	1	2	...do
W. W. Belden, Wisconsin	2	4	14	...do
James W. Easterly, New York and Missouri	11	28	118	...do
William H. Perry, Wisconsin	1	2	4	...do
D. F. Edwards, Illinois	11	8	811	...do
M. M. Harrison, Wisconsin	1	6	9	...do
L. W. Ingalls, Illinois	2	6	9	...do
William H. Waggoner, Illinois		6		...do
L. C. Chamberlin, Illinois	2	4	12	...do
William Lee, Ohio	2	7	7	...do
B. Geithmann, Illinois	5	27	43	...do
J. D. Willson, Illinois	1	2	2	...do
C. L. Kron, Iowa	1	2	4	...do
D. S. Zessin, Illinois	1	2	2	...do
L. W. Miller, Illinois	1	4	9	...do
George Westolp, Illinois	2	7	8	...do
J. Jeffs, Illinois	2	9	16	...do
Samuel Harrison, Wisconsin	11	16	110	...do
Torkel Keirson, Iowa	1	5	11	...do
John R. Caldwell, Illinois	1	2	22	Oregon
James B. McElhany, Oregon	2	11	58	...do
M. W. House, California	3	13	250	
Washington Farmer, California	4	120	700	
S. M. Farmer, Ohio	3	9	31	California
John Hale, Ohio	7	17	131	...do
Thomas Dickins, Ohio	1	5	21	...do
John Woods, Wisconsin	4	13	79	Oregon
James Wison, Wisconsin		1	3	...do
John McClure, Wisconsin	14	33	271	...do
Charles Bellnap, Wisconsin	9	17	139	...do
William Watson, Ohio	3		38	California
O. Sproul, Missouri	3	10	35	...do
A. Donovon, ———	1	4	13	...do
S. Zrivers, Indiana	7	13	33	...do
T. Zrivers, Indiana	9	17	108	...do
J. T. Wadsworth, Indiana	17	31	371	...do

STATEMENT—Continued.

Names and residence.	Number of wagons.	Number of persons.	Number of stock.	Place of destination.
Millard Robinson, Illinois	1	3	13	California
Levi H. Chasch, Illinois	7	11	47	do
G. K Campbell				
John Luos, Illinois	11	19	113	California
Joseph A. Jacobs, Minnesota				
J L. Criss, Minnesota		1	4	California
William H. Robinson, Illinois	10	21	62	do
N. S. Batcheller, Illinois	3	7	29	do
S H. Blonger, Minnesota	7	17	129	do
Ed N. Wadsworth, Indiana	27	42	27	do
R. V. Newsham, Minnesota				do
D. W. Anderson, Iowa	1	3	3	do
W. Roberts, Iowa	1	1	12	do
W. Y. Franel	1	4	10	do
A. Denniston, Iowa				do
H. Littell, Iowa	1	4	2	do
W. W. Waynick, Iowa	1	4	14	do
E. Culbutson, Iowa			14	do
H. J. Roderick, Iowa	1	3	4	do
Aaron Gilbert, Iowa	1	13	7	do
George Boardman, Iowa			3	do
Gilbert W. Hoffman, Iowa		2		do
James F. Stout, Iowa				do
Richard Alexander	1	3	8	do
Charley Switzer, Missouri				
William Alexander, Illinois	11	13	97	California
J. Collins, Illinois	1	2	4	do
R. H Stralburg	1	3	4	do
John C. Fink, Iowa	1	3	4	do
Elonzo Odell, Iowa				do
O. B. Dodd, Iowa				do
H. W. Briggs, Iowa	3	13	56	do
O. S. Howe, Maine	1	3	8	do
John W. Anderson, Iowa				do
John Stinger, Ohio				
Thaddeus Sterling, Illinois				
Jesse H. Chrisman, Pennsylvania	1	4	8	California
Joseph E. Rapp, Pennsylvania	1	4	8	do
James A. Field, Pennsylvania	1	4	8	do
Elijah Jacobs, Illinois	1	5	8	do
David Ashby, Illinois				
C. A. Trueman, Wisconsin	2	11	31	California
William Moon	1	2	6	do
N. B. Ingram, Iowa	6	24	36	do
Patrick Haugh, Iowa	1	2	20	do
Andrew McGee	2	6	12	do
Ivory McKinney	2	15	500	do
George W Gilbert	3	12	12	do
A. D. Buck, Wisconsin	6			
Th. Hale, Illinois	1	3	8	California
Thomas Parker, Illinois	1	5	8	do
Dawson Green, Illinois	1	6	8	do

STATEMENT—Continued.

Names and residence.	Number of wagons.	Number of persons.	Number of stock.	Place of destination.
Rufus C. Gates	1	3	8	California
Alfred Sutton	1	1	2	do
Luke Shaw	3	5	10	do
Charles Duncan, Illinois	1	5	8	do
J. Jenkins, Missouri	3	11	90	do
John H. Warrington, Iowa	1	5	8	do
Edwin Green, Illinois	1	6	8	do
Robert Witherspoon, Illinois	2	3		do
Ira Trelsher, Wisconsin	3	6	13	Oregon
John Thomas, Wisconsin	1	3	4	do
William Christy, Ohio	1	1	3	do
F. M. Scott, Missouri	1	3	7	
Andrew Clark, Iowa	2	10	19	Oregon
W. W. Markwell, Iowa	1	2	4	do
Edward F. Pearce, Wisconsin	6	24	67	California
Nathan Hall, Iowa	1	3	2	Oregon
Jason C. Pratt, Michigan	1	2	11	do
M. P. Scott, Missouri	1	3	12	do
James Ritchie, Maryland	1	3	6	do
Daniel Shipper	4	16	20	California
Fr. Bath				
G. W. Gallanar				
David Crock				
Henry Boughnow				
Nicholas Gallanar				
William Shiffer				
John F. Shiffer				
James M. Shiffer				
H. P. Hawkins				
Abraham Ede	2	9	80	California
John Walters	1	3	6	do
John Stewart	1	3	10	do
Robert Panter	11	5	4	Oregon
William Panter				
James Panter				
Samuel Slife, Iowa	4	10	18	California
John B. Collins, Iowa				
Richard Loney, Iowa				
Zebulon Walker, Iowa				
Sutler Walker, Iowa				
Abram Wilson, Iowa				
Ichabod Hubbard, Iowa				
A. C. Ely, Iowa				
Edwin S. Brown, Michigan				
D. S. Blouse, Iowa				
G. O. Vose, M. T.	4	14	22	California
C. D. Hapgood, M. T				
William N. Bulb, Minesota	5	19	33	California
Aaron Lampher, jr				
D. Berry, Iowa				California
William Morse, Iowa				

The undersigned, emigrants from Iowa and other States to California, desire to state, for the benefit of those who may emigrate hereafter, that they travelled the road leading by Salt Lake and found it very mountainous and rough, and most of the streams on said road were bridged and ferries established, over which exorbitant tolls were exacted for the passage of trains and teams; and where there were no bridges or ferries over the streams the fords were not only difficult but dangerous. They would also state, for the benefit of those who may emigrate hereafter, that they were compelled to pay from twenty-five cents to five cents per head a night for pasturage of their stock at Salt Lake and as far up as Bear river, a distance of nearly one hundred miles. That for about one hundred and twenty-five miles from the South Pass, towards Salt Lake City, the country was nearly destitute of grass, and might almost be called a barren waste, and the road strewn with carcasses and bones of dead animals lost the present and past seasons, caused doubtless by the great scarcity of grass; and they especially advise all future emigrants not to travel the Salt Lake road.

John E. Movers.
Z. N. Hewitt.
E. E. McAvoy.
G. A. Quick.
Joseph Stiffler.
David Davis.
Levi Adams.
Lewis Herren.
Hiram Young, Mercer county, Pennsylvania.
John Babehson.
Wm. Ostander, town of Winterset, Iowa.
Thomas Trester, Missouri.
J. C. Halloway, Honey Grove, Fannin county, Texas.
Mark Anthony, Indiana.
William Henry Ford, Illinois.
Duncan McKay, Vermont.
Charles Sullivan, Minnesota.
S. V. B. Shull, Kansas Territory.
William Peasly, his x mark, Minnesota.
S. H. Hartly, jr., Illinois.
John E. Janes, Minnesota.
Samuel Renslow, his x mark.
Fredrick M. Frisbee, Minnesota.
Samuel Ash Davidson, Illinois.
Joseph Jones, Indiana.
H. Reynolds, Minnesota.
William McIntosh, Minnesota.
Alex. Phillips, Arkansas.
L. M. Lawley, Arkansas.
A. J. Ruxby, Pulaski county, Arkansas.
Robert Rolston.
David Athy.
John Athy.
R. H. Bierly.

W. A. Townsend.
N. D. Townsend.
E. Mownsend.
S. C. Movers.
A. S. Lineback.
J. W. Taylor.
Wm. Dutton.
Philo. Clark.
John M. Chipman.
L. B. Trowbridge.
N. A. Trowbridge.
Charles Harner.
A. D. Miller, Michigan, (to California.)
A. A. Miller, Michigan.
Edward Bonyman, Illinois.
Daniel Lathrop, Vermont, (California.)
John Bonyman, Illinois, (California.)
H. F. Bennett.
William Haskell, Maine.
A. J. Young, Maine.
R. C. Brann, Maine.
A. C. Day, Boston, (bound for California.)
Sam. Haskell, Maine.
Enoch Philbrick, Maine.
Daniel Bryan, New York.
Egbert Hizrodt, New York.

Names and residence.	Number of wagons.	Number of persons.	Number of stock.	Place of destination.
Lucian Wright	4	19	42	California
Roswell Burt				
R. S. Walandtt	1	2	2	California
Asa Butler, Wisconsin	1	2	4	do
F. Carnsworth, Wisconsin	2	4	8	do
G. W. Reynolds, Wisconsin	1	4	2	do
M. H. Balsic, Wisconsin	1	5	6	do
E. Grenard, Indiana	2	3	9	do
R. Anderson, Illinois	1	4	9	do
Horace Douglas, M. D., Michigan	1	2	2	do
R. B. Ware, Pennsylvania	2	5	17	do
Jacob Christian, Illinois	1	8	10	do
Jacob R Vogdes, Illinois	1	4	12	do
Tobias Teller, Illinois	2	3	9	do
D. W. Rinewalt, Illinois	1	4	10	do
William H. Freid	2	7	8	do
J. T. Miller, Illinois	1			do
Charles Shrom, jr	4	1	20	do
Robert R. Miller	1	2	4	do
A. H. Simpson, Illinois	1	3	2	do

STATEMENT—Continued.

Names and residence.	Number of wagons.	Number of persons.	Number of stock.	Place of destination.
Robert Orfield, Illinois	1	3	5	California
Ogden Edwards, New York	2	9	8	...do
Isack G. Cork, Pennsylvania	1	4	4	...do
John Baird, New York	7	24	14	...do
Alexander Brown, New Hampshire	6	26	28	...do
A. F. Brown, New York	....	....	....	Oregon
Jas. W. Maxwell, Illinois	5	12	29	...do
Hiram Stuart, Nova Scotia	1	2	5	...do
Jacob Elliott	2	11	12	...do
Henry Emrick, Iowa	2	14	21	...do
William Emrick, Iowa	1	2	4	...do
R. F. Lane, Missouri	6	26	610	....
Isaac Pferheimer, Missouri	6	26	17	....
Thomas Lane, Missouri	11	29	78	California
W. A. Huff, Missouri	12	20	1,010	...do
John T. McFarlan	1	4	28	...do
Solomon	....	2	6	...do
John Lamb, Missouri	4	12	60	...do
T. J. Faulkner, Missouri	1	6	19	...do
R. V. Kelly, Missouri	5	24	822	...do
[Unintelligible]	4	19	89	...do
W. H. Wise, Illinois	2	4	12	...do
C. H. Bingham, Wisconsin	1	3	12	...do
B. M. Rabert, Missouri	7	18	79	...do
J. W. Lambert	2	4	11	...do
John Deasy	1	1	3	...do
Zina H. Fairchild	4	11	16	...do
A. Evans, California	5	21	325	...do
R. E. Ross, California	1	3	4	...do
W. Smith, Ohio	1	3	22	...do
Isaac Harp, Illinois	2	7	11	...do
J. N. Evans, Ohio	4	20	300	...do
Henry Cosgrave, Ohio	....	1	....	...do
M. L. Crawford, Iowa	....	1	....	...do
Sam. H. Dewey, Iowa	....	1	....	...do
Peter Gnio, Michigan	2	14	65	...do
Frank House, Ohio	9	28	111	...do
John Brown, Iowa	6	23	36	...do
John Dobkins, Iowa	2	10	29	...do
Charles Gilbert	8	23	113	...do
Zalbrah Archibald	1	2	21	...do
U. S. Ingram, Council Bluffs, Iowa	1	2	2	...do

www.ingramcontent.com/pod-product-compliance
Lightning Source LLC
LaVergne TN
LVHW011140110826
845150LV00008B/2430

* 9 7 8 1 4 1 8 1 9 3 2 6 3 *